CA

D1513765

Richmond upon Thames Libraries

Renew online at www.richmond.gov.uk/libraries

LONDON BOROUGH OF
RICHMOND UPON THAMES

90710 000 467 551

Plant Trees Sow Seeds Save The Bees

Simple Ways To Bee-Friendly

NICOLA BRADBEAR

Published in 2021 by Witness Books, an imprint of Ebury Publishing
20 Vauxhall Bridge Road,
London SW1V 2SA

Witness Books is part of the Penguin Random House group of companies
whose addresses can be found at global.penguinrandomhouse.com

First published by Witness Books in 2021

www.penguin.co.uk

A CIP catalogue record for this book
is available from the British Library

ISBN 9781529108774

Colour origination by Altaimage Ltd, London
Typeset by @seagulls.net
Printed and bound in Great Britain by Clays Ltd, Elcograf S.p.A.

The authorised representative in the EEA is Penguin Random
House Ireland, Morrison Chambers, 32 Nassau Street,
Dublin D02 YH68

MIX
Paper from
responsible sources
FSC
www.fsc.org   FSC® C018179

CONTENTS

CONTENTS

FOREWORD

Bugs. Creepy crawlies. Pests.

Insects are perhaps the least loved and most underappreciated members of the animal kingdom. In fact, we humans spend quite a lot of time and money on repellents and insecticides trying to banish insects from our lives – and we've been quite successful.

A recent study found that more than 40% of insect species are declining and a third are endangered. The total mass of insects is falling by 2.5% a year, according to the data, and if that continues, within a century they could all be gone. But given that we don't love insects very much, would that be a bad thing?

Well, yes. It would be a very bad thing, because without insects, many of the things we do love – trees, flowers, birds and a whole host of cuddly, furry, adorable animals wouldn't exist at all. And

in fact, neither would we. By helping bring about
the end of insects, we are speeding up our own
extinction. Because insects are both food for a
host of other species, and pollinators of the plants
that are eaten by everything from elephants to
dormice, orangutans to us humans.

Like it or not, we need insects, and for all of us
to thrive we need to treat them, not as foes, but
as the much-needed friends and allies they are.
Luckily, our insects have a champion. The author
of this book, Nicola Bradbear, has made it her
life's work to educate communities all around the
world on the importance of insects, particularly
stripey ones. As founder and director of the
charity Bees for Development, she has worked
as an adviser and consultant all over the world.

In Monmouthshire, where she lives and where
the charity is based, Nicola helped to establish
the Bee Friendly Monmouthshire campaign group
to address the dramatic decline in numbers of
honey bees, bumblebees and other pollinators
such as wasps and hoverflies, moths and
butterflies.

Visit the county and you will notice that many of the verges, roundabouts and other small parcels of land have been planted with wild flowers to provide food and habitat for these vital insects. You will also notice how beautiful it is to have wild flowers along our roadsides. We benefit as much as the insects.

By buying this book, you can start to get to know some of our stripey insect friends and to appreciate how much they enrich so many aspects of our lives. But I also hope that you will be inspired to become an insect champion and encourage others to do the same, because the rewards will be many, and it will definitely give you a bit of a buzz . . . !

Kate Humble

INTRODUCTION: WHY DO WE NEED INSECTS?

The plant life on our planet flourishes thanks to the labour of insects. From dawn to dusk, hour after hour, day after day, our insects are ferrying pollen from one flower to another. This work enables flowering plants to create their seeds, to bear fruit, and means that future generations of plants will continue to thrive.

Most of the fruit, vegetables and crops that we rely on for food, and that feed animals, too, rely on this busy process.

Next time you step outside, look for some flowers – any flowers, anywhere. If they are real (not plastic or nylon!), you should find some insects feeding on them. That is why those flowers exist – to attract insects to the plant, and no other reason. The insects are feeding, and at the same time they are pollinating the plant, enabling it to produce seed for the next generation of plants, to feed the next generation of bees, and everything else.

Once you begin to notice them, you'll see bees and other stripey insects constantly searching out every useful flower.

Every scrap of earth can support a few flowers, which will in turn support a few insects, which will pollinate more plants and support a few birds, too. Flowers and insects are at the beginning of long food chains that support biodiversity, and right now the world's capacity

to produce food is being undermined by our
failure to protect them.

Stripey insects, including bees and many other
insects (and not just the stripey ones), are incredibly
important for continued life on Earth. We know that
they pollinate our food crops, but that is just one of
the crucial services that they carry out for us.

- Insects are a vital part of the food chain:
  they are the food that supports populations
  of larger animals – birds, bats and other
  insect-eaters, like amphibians and fish.

- They work to keep everything decomposing
  nicely, eating waste and unlocking nutrients
  to keep the natural cycle going.

- Insects are important pest controllers – for
  example, wasps, hoverflies and ladybirds
  feed on aphids and greenfly. They are the
  gardener's friends.

- They are important soil engineers – for
  example, ants keep the soil aerated and
  healthy for plants to grow.

3

- Creatures of beauty – insects moving from flower to flower are wonderful to see and bring us joy. It is hard to imagine that joy ending completely, so we must do all we can to preserve them.

Depending on where you live, you will have your own range of stripey insects. The only one that exists almost everywhere on Earth and that you are highly likely to see is the honey bee, which people have transported from where it lives naturally to almost every nation. The European wasp and some bumblebee species have been transported by us outside their natural distribution range, too – for example, the European wasp has been common in New Zealand since the 1940s. Otherwise, you will see your own local species of bees, wasps and hoverflies that are similar to, but different from, those illustrated in this book. If you live in a temperate climate you will see bumblebees – although they do not occur naturally in Southern Africa, Australia or New Zealand, where you have other special bees occupying their niche. For example, in Australia you will see amazing blue-banded bees, carpenter bees living in dead wood, and many other social and solitary species. North America has its own species of shiny blue-black carpenter bees, and many species of mining bees and leaf-cutter bees – the 'same but different' from those described in this book.

If you are in a tropical climate anywhere on Earth – except for desert areas with no flowers – you will see stingless bees of a huge variety of species; some large like honey bees and others tiny, almost like mosquitoes. These exist all around the tropical world, and there are at least 500 species. These bees store honey as do honey bees, but they cannot sting. They occur in tropical regions of Asia, Australia, Africa and the Americas. Although they don't sting, they are able to defend themselves with their bite. Stingless bee honey is extremely highly valued for its medicinal properties: it is antimicrobial, and has been used by indigenous people in tropical regions to treat eye problems.

The bees in this book all occur in my garden, here in Wales, in the UK. The ones that you will find where you live will be similar, though not exactly the same species. The aim of this book is to help you to discover how many amazing, different stripeys are sharing your spot on the planet. It is of no matter whether you can put the right scientific name to each – almost nobody can do that – the important point is to be aware of them, and to help them to survive.

GETTING TECHNICAL – WHAT ARE BIODIVERSITY AND BIOABUNDANCE?

Biodiversity combines the terms *biological* and *diversity* and means all the types of life on Earth – the plants, animals, fungi and micro-organisms – all the communities they form and the habitats in which they live. Put simply, the more types of plants there are, the more types of insects, birds and mammals you will find, too.

The opposite of biodiversity is monoculture.

We need plant life in abundance to make it possible for our wildlife, and crucially our insect populations, to survive. We do not just want biodiversity within a few manicured nature reserves, we need to restore populations of insects by using every scrap of earth to restore nature's bioabundance. Roadside verges and hedgerows provide important corridors for wildlife, connecting all the natural areas. Every scrap of wasteland offers scope to restore the bioabundance that nature needs to survive and to thrive.

Gardens are a significant natural resource: a tapestry of gardens across nations offer a hugely important refuge for bees and other stripey insect friends. The fact is that ALL land birds depend on insects at some time in their lives, including seed-eaters. For example House sparrows are seed eaters when they are adult, however they cannot survive as nestlings without being fed on insects. Insects come into all food chains.

It's a simple equation: no flowers = no insects = no birds.

POLLINATION AND POLLINATORS

For life on Earth to continue, plants need to produce seed to create the next generation of plants. However, the problem for plants is that they are rooted to the spot where they are growing, so how do plants procreate?

Millions of years of evolution have resulted in a close, mutually beneficial relationship between plants and bees: plants depend upon bees to transfer pollen grains (tiny packets of genetic material) from the male part of one flower to the pollen-receiving, female part of another flower. This allows fertilisation to take place, and a new seed begins to develop inside the female flower. In return for this service, and to ensure that there are future generations of bees to continue the pollination of plants, the plants feed the bees with nectar and pollen.

NO FLOWERING PLANTS = NO FUTURE GENERATIONS OF BEES
(OR ANYTHING ELSE)

NO BEES = NO FUTURE GENERATIONS
OF FLOWERING PLANTS (OR
ANYTHING ELSE)

And if there are no flowering plants there will be
no seeds or fruits for us to eat. Imagine that – no
coffee, no chocolate, no oranges, no apples, no
fruits and seeds, and no life on Earth.

Much of the food we eat starts as a flowering
plant, and even the clothing we wear – all
the cotton and linen clothes – originate from
flowering plants.

And, of course, fruits and seeds feed and nourish
the rest of life on Earth, too, because they support
biodiversity (see page 7).

Flowering plants began to evolve about 130
million years ago. Before that there were just
green ferns and evergreen conifer trees. When
flowering plants, and their associated pollinating

insects, began to evolve, they rapidly flourished, supplying the abundance of flowers, fruits and seeds that we, and much of nature, depend upon for food.

A pollinator is an insect or something else that transfers pollen from one plant to another of the same plant species. Bees are very important pollinators; however, other insect pollinators are available! In fact, many species of insect function as pollinators – for example, hoverflies, beetles, butterflies and wasps.

Some flowering plants depend on birds and small mammals to bring about pollination. For example, birds pollinate tropical flowers, often flamboyant red ones used by florists – like the well-known *Crocosmia* and *Strelitzia*, the bird of paradise flower, which in their native South Africa are both pollinated by the sunbird, whose long, curved bill perfectly fits the long red tubes of these flowers. Flowers attract birds more through their colour than their scent, and so these tropical bird-pollinated flowers tend to be red, pink, yellow and orange, because these are colours that birds see well. The extended tubular blooms can hold a great reservoir of nectar to accommodate birds' appetites, and to keep them engaged and picking up the flower's pollen for longer periods.

WHY DO INSECTS NEED OUR HELP?

We believe that insect numbers have reduced by something like 95 per cent during the past fifty years, although because nobody anticipated their decline, we do not have much historical data from anywhere in the world on insect abundance. Accurate data does come from sixty-three nature reserves in Germany, which reveal a 76 per cent drop in insect abundance during the twenty-seven years between 1989 and 2016. And this rapid decline is in protected, natural areas!

Insect decline has been, of course, far greater in agricultural and industrialised areas.

Surely this must alert us to the problems facing nature: the impact of land development for roads, industry and housing, of intensive agriculture and climate change? And it's not just insects: plants, birds, fish and mammals are all in decline. If we were coal-miners, we'd be up to our eyes in dead canaries by now.

The main reasons why insect numbers have reduced dramatically are the absence of their food plants and the presence of pesticides. There are other reasons, too, including introductions of new species, pests and viruses, and also climate change.

Insects have been on Earth since long before we appeared, and in many ways, they create the world in which we live. Can you imagine a world without flowering plants? Much of your breakfast, lunch and dinner would not exist.

The good news is that you can help turn these stats around with your everyday choices, with

your voice and vote, with your spade and, if
you can, with your money, too. Everyone doing
a little adds up to a big difference. And when
we get things right for insects, we begin to get
things right for everything else, too. Even one
flowering lavender plant on a windowsill makes
a difference, providing a few stripeys with their
essential food: the pollen and nectar in flowers.

We are only just learning about the sophistication
of insect life. Even tiny hoverflies can migrate
hundreds of miles, and your garden might be the
spot where they touch land in urgent need of a
sip of nectar.

All of the proposals in this book will help stripey
insects and make your life richer and more
interesting, too. Most of these proposals are
no-cost, or low-cost, yet by doing them you will
make a significant difference to life on Earth.
This is because insects cannot live only on nature
reserves; they need corridors of food plants to
enable them to move from one place to another,
to find food and nesting sites, and to allow
populations to survive in prevailing conditions.

Insects are tiny, and each one needs only a tiny scrap of food: every dandelion provides food for a few more insects to survive and thrive.

Restoring habitat for insects is a relatively inexpensive way for everyone to help recover the Earth's biodiversity, something that we have trashed during the past few human generations. One problem is that we cannot readily see declining nature; we don't see starving insects and diminishing populations, and we do not miss what we didn't know existed.

People born in the fifties and sixties will remember a time when insect life was far more abundant, when car windscreens and lights became covered in insects on summer evenings. However, for younger generations, an insect-free life has become the new normal. You might think it sounds more pleasant to have fewer insects, but this situation has disastrous consequences for the future of our planet.

GETTING TO KNOW STRIPEYS

This book will help you get to know some of
the stripey insects that live alongside you. We're
focusing on stripey ones here because they are
the easiest to spot!

It doesn't really matter if you can assign the right
name to the right stripey insect, the important point
is to be aware of their existence, and to give them
food and safe lodging. A suburban garden with
even a modest mix of flowers and habitats should
be home to hundreds of different insect species.

I promise you that when you begin to start looking, you will be able to differentiate between our stripey friends. You will soon realise which insects favour which flowers. And when you begin to recognise a few insects, this will add a whole new dimension to your life – you might even find insect-spotting to be an addictive pastime! A large bed of lavender, like those you often see in municipal plantings, is alive with a great range of stripey insects – try to notice how many different species are enjoying the lavender-fragranced feast.

DON'T BE AFRAID OF STRIPES

Wasps have dramatic yellow and black stripes, and this pattern and colour combination say 'wasp – watch out!' to everyone. Throughout the animal world, yellow and black stripes signal danger. We learn from an early age that wasps sting and so we fear them, even if we have never been stung by one. Yellow and black stripes have become our internationally recognised symbol for danger – notice the familiar pattern used on sticky hazard tape, to indicate steps and obstacles, and on road signs and vehicles.

Wearing black and yellow stripes is a clever way for a creature to protect itself from predators: many completely harmless insects that cannot sting, such as hoverflies, protect themselves by wearing the wasp's yellow and black stripes. British naturalist Henry Bates wrote about this concept in 1861, while exploring the Amazon rainforest, and called it Batesian mimicry.

However, the downside to this fear and anxiety is that humans are too often terrified of highly beneficial, and often harmless, insects.

Of course, female honey bees look a little bit like wasps, and honey bees can sting. Müllerian mimicry is the name given to describe the situation where two potentially stinging species evolve to look similar to each other. This is because there is efficiency to be gained by predators looking alike: when this happens with two harmful species, together they have a greater chance of successfully repelling predators.

There is also another reason for bearing stripes, which is that a fast-moving stripey object is harder to spot than something solid-coloured. Does this 'flicker fusion effect' also explain the familiar stripey football 'strip'? Zebras are perfectly camouflaged for the African savannah grasslands, where they need to avoid lions and other predators that would find it easier to spot solid black- or white-coloured prey.

INSECT SOCIETIES

We categorise species of bees and wasps
according to whether they show social or solitary
behaviour: both categories have behaviours and
a capacity for learning that we are only just
beginning to appreciate. Social insects like honey
bees live as a 'superorganism' and have evolved
complex behaviour in which individual members
undertake specific tasks to maintain the society,
which they achieve by means of communication.
We have learned most about honey bees because
we have enjoyed a very close relationship with
them for thousands of years through beekeeping.
Recent research reveals the amazing ways in
which honey bees communicate and reach
consensus to make optimum use of available
resources.

One social bee cannot live on her or his own –
she or he can survive only as part of a bee
colony. In the case of honey bees, the colony

consists of thousands of (female) worker bees, hundreds of drone (male) bees and one queen (female) bee. This bee colony is not really 'a family' – although this term is often used – because the whole colony is just one individual entity. It is more like each individual bee is one cell of a larger animal. However, that analogy does not work well either, as each bee is of course made up of cells, too. Therefore we use the term superorganism, which is a social unit made up from individuals (which could not survive alone), within which there is division of labour, with each individual working for the good of the whole. The honey bee colony is no longer regarded as an autocratic monarchy run by the queen bee, rather, the colony reaches democratic decisions by consensus, based on data arriving from many different routes. For example, when a honey bee colony plans to swarm (with the purpose of reproducing), scout bees identify available nesting sites in the vicinity and communicate this information to the bees back home, who consider all incoming data and select the new site that receives the most votes.

MEET YOUR STRIPEY NEIGHBOURS

Stripey insects include bees that store large volumes of honey – ones that can sting, as well as stingless bees that do not – in addition to bumblebees, solitary bees, wasps and hoverflies. Although they all look superficially similar, you can tell them apart and find out more about their different lifestyles. Read on to learn how to begin to differentiate the various stripeys that you can find.

THE HONEY BEE

Learn first to recognise the most common stripey insect, which can live only as part of a large colony – the superorganism.

Social insect no. 1: the honey bee

The remarkable thing about a honey bee colony is that it is a permanent organism, surviving from year to year through long periods when there are no flowers, by storing honey.

Honey is concentrated nectar; it is collected by bees from flowers, then dehydrated by them to create the honey. It is this feature that has made honey bees attractive to humans throughout our history, and continues today.

All this work of collecting nectar and reducing its water content to make and store significant volumes of honey means that honey bee colonies

must be large – at the height of summer there may be 50,000 or more bees in one colony.

Honey bees naturally live inside tree cavities and inside beekeepers' hives, too. The species illustrated here is known as *Apis mellifera* and occurs naturally from north of the Arctic Circle, through all of Europe and the Middle East, and all of Africa. This bee has been introduced worldwide, and you will find it in gardens everywhere. There are many different races, or subspecies, of this bee – with different characteristics that enable them to survive across widely varying climates, from −20°C (−4°F) in European winter, to 40°C (104°F) in the Middle East. Nowadays, this species has been introduced to every nation and is utilised in commercial beekeeping: you are highly likely to find honey bees foraging on flowering plants everywhere. They are more slender than a bumblebee, and brown in colour, although varying from dark yellow to brown to black. Compared with bumblebees they have short tongues, so you will see them foraging on white clover but not red clover, nor plants like *Buddleia*, for which the longer tongues of some bumblebees and butterflies are needed.

SEVEN BUMBLEBEES

The table on the next page shows seven common bumblebees – notice their tongue length and the food plants on which you are likely to find them foraging. Noticing the varieties of plants that different bee species tend to favour is the best way to begin to discriminate between different species. **Tongue length** is the important factor determining which plants different bumblebee species feed upon.

SEVEN BUMBLEBEE SPECIES COMPARED			
	EARLY	BUFF-TAILED	WHITE-TAILED
TAIL	SMALL, GINGERY RED	WHITE WITH BUFF-ORANGE HAIRS	WHITE
THORAX	1 YELLOW BAND MALES HAVE YELLOW FACE	1 BROAD, DARK YELLOW BAND	1 PALE YELLOW BAND ON FEMALES 2 PALE YELLOW BANDS ON MALES
ABDOMEN	1 YELLOW BAND, OFTEN FAINT	1 BROAD, DARK YELLOW BAND	1 YELLOW BAND
TONGUE LENGTH	SHORT	SHORT	SHORT
TYPICAL FOOD PLANTS	*ECHINACEA, GAILLARDIA, GERANIUM,* HELLEBORE, ROSEBAY WILLOWHERB, WHITE CLOVER, WHITE DEAD-NETTLE, WELSH POPPY		

Social bumblebees, like honey bees, live together
in one colony, and they also store honey.
However, the whole colony does not persist from
one year to another like that of honey bees, and

TREE	RED-TAILED	COMMON CARDER	GARDEN
WHITE	BRIGHT RED	GINGER	WHITE
GINGER BROWN	BLACK – FEMALES ARE ALL BLACK, MALES HAVE 2 YELLOW BANDS AND YELLOW ON FACE	YELLOW-BROWN	2 YELLOW BANDS
BLACK	BLACK	GINGER AND PALE BLACK BANDS	1 YELLOW BAND
SHORT	MEDIUM	LONG	VERY LONG
	ARTICHOKE, CHIVES, *HEBE*	*HEUCHERA, HELIANTHUS, VERBENA*	*AQUILEGIA, BUDDLEIA,* FOXGLOVE, RED CLOVER

therefore they do not need to store large volumes of honey, as honey bees do. Each bumblebee colony consists of a few hundred individual bees, as opposed to the thousands in a honey bee colony. At the end of the season, all of the bumblebees will die except for the queen, who will hibernate alone over the winter.

Worldwide, there are about 270 species of bumblebees, but they do not occur naturally in Australasia, Central or Southern Africa or the Indian subcontinent. The smaller size of bumblebee colonies means that they are more sensitive to pesticides than honey bee colonies are.

Once you begin looking at bumblebees you will realise that you can indeed differentiate between some of them. The first thing to notice is the tail: is it white, is it orange-red, or is it the same colour as the rest of the bee?

Next, look at the banding pattern on the bee. Many bumblebees have a white tail with yellow and black stripes across the rest of their body. Look more carefully and you will notice that the number of these stripes varies. A good idea is to take a picture on your phone – it won't be perfect, but it will give you a chance to look closely at the bee in your own time (see page 147).

The size of the bee does not help much with identification, as queens and workers (females) and the drones (males) are different sizes, and the size of successive generations can vary during the season, too: the first workers are small as they are nourished only by the early spring food that their mother – the queen – could find on her own. Bees developing later are larger as they are fed by all their sisters, when there are more flowers available. Nature is wonderful!

However, the noticeably big bumbles that you see early in spring are queens that have recently emerged from their winter hibernation, seeking a place to establish their nest and begin their family. Right now each queen has a critical job to find enough early spring food to build up her strength, develop her ovaries and lay eggs, then look after them until they hatch – her first daughters, who must grow and mature until they are able to take over food foraging to support the family. For the rest of her life, the queen will stay inside the nest, laying eggs to develop into more females (workers) and males (drones).

WHY HONEY BEE IS TWO WORDS AND BUMBLEBEE IS ONE WORD

This is not going to change your life, it's just a nice fact to know! A honey bee is a type of bee and a house fly is a type of fly, so 'honey bee' and 'house fly' are two words. However, when we come to bumblebee and butterfly, these are common names – there is no type of bee named bumble and no type of fly named butter (and neither is it a fly). So in these cases it's just one word.

Red-tailed bumblebee
Bombus lapidarius

This bumblebee is the easiest to identify – the females are black with a distinctive red-orange-ginger tail, while the males also have two yellow bands on their thorax and some yellow on their faces. Its Latin name is *Bombus lapidarius*, and this bumblebee is present throughout Europe, where you will find it flying during spring and summer months. This bee has a medium-length tongue (6mm/¼in) and you will often see it foraging on dandelions, thistles, knapweed and daisy-type flowers.

Buff-tailed bumblebee
Bombus terrestris

Buff-tailed bumblebee queens emerge from
their hibernation in early spring and search for
somewhere to begin their nest: a welcome harbinger
of spring ahead. This common (in Europe) bumble
has a white tail with a narrow ginger border and
two broad yellow bands: one on the thorax and
another on the abdomen. The narrow ginger border
to the white tail separates this species from the
white-tailed bumble (overleaf). It has a relatively
short tongue, so therefore you will see it foraging
on flowers with short corollas like dandelion, white
clover, cotoneaster and many other garden flowers.
You will often see it foraging alongside honey bees.

This species is widely used in commercial
horticulture to pollinate crops growing in
polytunnels – for example, tomatoes and
strawberries. This has resulted in this bumblebee
now being exported to and living in places far
beyond its native range, such as Argentina, Chile
and Japan. This is one of the ways that we spread
bee diseases and viruses around the world and
cause great harm to native bee populations.

White-tailed bumblebee
Bombus lucorum

This bumble emerges in early spring. You can differentiate it from the buff-tailed bumble (previous page) because this bee always has a pure white tail with no orange/yellow banding at the edge of it. The rest of the banding on the white-tailed bumblebee is pale yellow.

Garden bumblebee
Bombus hortorum

This is a large European bumblebee seen from
late spring onwards. It has a long black face
with a long tongue and seems to make a louder
and lower buzzing noise than the others. It has
three yellow bands (two on the thorax, one on the
abdomen) and a white tail. This is the longest-
tongued bumblebee (indeed, it has the longest
tongue of any bumblebee, worldwide!), so during
spring you will see it, and hear it, foraging on
flowers like *Aquilegia* (page 103). These flowers
have an exceptionally long corolla, so the nectar
is accessible only to suitably long-tongued
insects, like the garden bumblebee. You will see
it foraging also on other long-tubed flowers, like
those of bluebells, borage, *Buddleia*, comfrey,
foxglove, lavender and red clover.

The garden bumblebee is similar to the buff-
tailed; you can tell them apart by counting the
yellow bands: the buff-tailed has two, the garden
bumblebee has three.

Tree bumblebee

Bombus hypnorum

This one is easy to find: a large, furry bumblebee with no yellow bands that you can hear buzzing as it forages on flowers. It has a ginger-coloured thorax, a black abdomen and a white tail. It is widespread across Europe and mainland Asia. This species likes to nest well above ground, often choosing bird boxes and roof cavities. The behaviour of the drones (males) is eye-catching: they display a special 'dancing' behaviour (bobbing about) outside nest entrances. This is 'nest surveillance' – often 25 or 30 drones hoping to spot a virgin queen to mate with. The worried householder need not be overly concerned, as tree bumblebee colonies only persist during late spring and summer, after which all will die off, except for the queen, who will overwinter in a tiny space somewhere, hopefully safe, until she emerges next spring.

Tree bumbles have short tongues and you will find them foraging on open flowers like bramble and single roses, clematis and dahlias.

Common carder bee

Bombus pascuorum

This is another easy-to-identify bee, as it is uniformly coloured with an orange thorax and stripey orange and black abdomen. It most resembles a tree bumblebee, but it is easily distinguished by the lack of white tail. The carder gains its name from its behaviour of gathering moss to cover its nest. It has a long tongue, so you will find it foraging on flowers with long tubes like foxgloves, thistles, viper's bugloss, red clover and bean flowers.

Early bumblebee

Bombus pratorum

As its name implies, queens of this bumble
species are on the wing early in the year, when
you may find them foraging on snowdrops,
crocus and *Pulmonaria* in late winter and early
spring. This is another red-tailed bee, but it also
has two yellow bands, distinguishing it from the
female red-tailed bumble, which has no yellow
bands, and the male red-tailed, which has two
yellow bands. It has a short tongue and forages
on white clover, borage, brambles and
raspberries. In the UK, this is the smallest
bumblebee.

A SOCIAL WASP

Wasps are a natural alternative to chemical pesticides because they eat aphids. They also pollinate plants and play their role in maintaining biodiversity as food for amphibians, birds and mammals.

Wasps deserve to be recognised for what they are: the gardener's friend. Yet almost every day someone tells me, 'I love bees, but I hate wasps.' Why is that? A wasp sting is certainly no worse than that of a honey bee. It seems to be deeply embedded within us that wasps are bad and bees are good. I would like you to consider that they are *all good*, and we need them. Using wasp traps is anathema to anyone who understands the value and role of wasps.

Wasps are important pollinators; as Darwin himself observed, helleborine orchids are entirely dependent upon them. Wasps and bees share

ancestors: bees separated from wasps when flowering plants began to evolve. While bees are vegan – they eat only plant products (pollen and nectar) – wasps are carnivorous and are predators who must catch insect prey (mostly aphids and flies) to feed the developing larvae back in their nest. And in view of the wasp's carnivorous leanings, the helleborine orchids attract them with a more 'meaty' scent!

As the summer progresses, wasp colonies have no more young to feed. The adult, mated queens must find a safe nook somewhere to hibernate until next spring. The rest of her family – all the adult workers and drones – will die by autumn and cannot exist over the winter. In late summer, these adult wasps can survive mainly on a carbohydrate diet as they just need fuel for flying, and thus they seek sugar wherever they can find it, foraging on flowers for nectar. Remember, they are predators that need to hunt for their food, so if they come across a jar of jam or a sugary drink the hungry wasp would like to take a sip. Is it this annoying behaviour that earns them their bad reputation? Wasps are not going to sting unprovoked, but they are likely to do so if you tread on them with bare feet or accidentally squeeze or threaten them in some way.

If wasps are disturbing you during your picnic the worst thing you can do is to flail your arms around – that looks like incitement to attack! Whatever is attracting them is giving off sweet and sugary scents, so put a lid on the jam or cover up sweet drinks. Now you know why Victorians had all those lacy things to cover food at picnics!

Like all of the stripey insects (except honey bees), only the queen wasp overwinters and she must begin her new colony when spring arrives. When she emerges from hibernation in spring, she's on her own until she raises some offspring. She begins by building a small starter nest – about the size of a golf ball – that is suspended in a cavity somewhere and has a dozen or so hexagonal cells in which to start the family. She lays an egg in each cell, and when they hatch, she must feed the growing larvae, cradling them at night to keep them warm. Building the nest and finding enough food for herself and her young is challenging for this single mother, and very few make it through this stage. It's about a month before the first young are able to join their mother in hunting for food, and now a new nest must be built to house the growing family. Wasps build their nests from a type of paper that they make by gnawing fibres from bark, timber, stumps, logs or wooden garden furniture. The fibres are mixed with saliva, then the wasp carries the ball of waspy *papier-mâché* back to its nest and incorporates it into the developing nest structure, which is made up of hexagonal cells, as with a bee's nest. Wasp nests are often located in the ground or in cavities in buildings or trees and are constantly readjusted as the colony increases in size. At the end of the season, as all the wasps die off except the queens, the temporary paper structures are left empty and gradually decompose.

FOUR SOLITARY BEES

Most bees do not live socially in large colonies and these are named 'solitary'. Most bees are solitary species; they live on their own, feed only themselves, mate, then create a nest in which the female bee lays eggs and provides food for next year's generation. Have you noticed any of them? Almost nobody can identify them all, but everyone can learn to recognise one or two. Next year, the adult females hatch and live for a short time during spring or summer, mate, then lay eggs for the following year's bees, provisioning the nest with food for the emerging larvae, protecting the tiny nest with clay (mason bees) or with leaves (leaf-cutter bees). The idea of the bee hotel is to provide a safe place for the eggs of tube-nesting bees to overwinter (see page 124 for how to make bee hotels).

Mason bees

Some of the first solitary bees to appear in spring
are the mason bees, and you will often see
these bees feeding on dandelions, which are a
wonderful source of nectar and pollen for these
early-flying bees. Commonly seen in Europe is
the red mason bee, *Osmia bicornis*, which is
about the same size but slightly more delicate-
looking than a honey bee and with a beautiful,
red, furry abdomen.

Wool carder bees

Anthidium manicatum

Look out for the garden plant lamb's ears
(*Stachys byzantina*), which gets its name from its
downy white leaves. The wool carder bee collects
this white down to line its nest – turn the lamb's
ear leaf over and you will see the tell-tale bee
scratches underneath. Look a little longer in the
nearby area and you are likely to discover a male
wool carder bee, fiercely deterring any other
insects straying into his territory surrounding the
lamb's ear plant.

Mining bees

These all nest in soil: you will see a small
'volcano' mound of soil with a tiny hole in the
top. Try not to damage these nests for the few
weeks while the bees are active. Worldwide,
there are thousands of species of mining bees.
Two that you are very likely to see on the wing
during spring in Europe, and which are easily
recognised, are the tawny mining bee *Andrena
fulva* (opposite) and the ashy mining bee
Andrena cineraria, which is a small black bee
with two broad, ash-grey bands across the thorax.

Late in the year when the ivy is flowering, look for
ivy bees (*Colletes hederae*), another mining bee
that nests in soil. Find some flowering ivy and there
could be ivy bees, honey bees and maybe some
late bumblebees feeding on these last of the year's
flowers – you may hear the throng of bees before
you see them. These ivy bees are active during only
these 4–6 weeks in autumn. They are solitary bees,
but because they sometimes nest in aggregations of
hundreds of nests, people mistake them for a 'swarm
of honey bees', and all too often they then kill them.

Leaf-cutter bees
Megachile centuncularis

There are many species of leaf-cutter bees and
you may detect them first by the characteristic
shape of their work, cutting oval linings for their
nests from the leaves of roses, rosebay willowherb,
lilac, *Leycesteria*, birch, ash and horse chestnut
trees, lilacs and honeysuckles. The bees
themselves look like slender honey bees; however,
they carry pollen not on the hind legs but on
the orange pollen brush located underneath their
abdomen – a distinguishing feature. These bees,
and their delicate nests, are a wonder to behold.
Please don't let any of your rose-loving gardener
friends consider them to be 'pests' and kill them
with pesticides – it is unbelievable but true that
some do!

FOUR HOVERFLIES

These are certainly the gardener's friend,
because they are important pollinators of plants,
and their larvae feed voraciously on aphids,
too. Everyone mistakes hoverflies for bees or
wasps, and indeed that is what evolution has
intended, but in fact hoverflies have no sting
and are totally harmless. Many hoverflies have
black and yellow markings to protect them from
predators. Remember this easy rule for stripey
insects: if it has two pairs of wings it is a bee or
wasp, if it has one pair of wings, it's a fly. And,
of course, the giveaway behaviour: hoverflies
hover near to flowers in a wonderful feat of
aeronautics.

There are around 300 species of hoverflies in
Britain, and most have only Latin names. Here
we introduce you to just four common ones that
you will find easily. Once you begin to look, you
will see that it is very easy to separate hoverflies

from bees, and that there are many different species of hoverflies in every garden.

Marmalade hoverfly

Episyrphus balteatus

You guessed it – this hoverfly has bright orange
stripes! This is one of the most common hoverflies
in Europe, and you will find it easily if you look
on flowering plants at almost any time of year.
The larvae feed on aphids, and the adults feed on
flower nectar and pollen. This hoverfly provides
valuable control of aphids in commercial
agriculture, although all are killed when crops
are treated with insecticide.

Hoverfly or common drone fly

Eristalis tenax

This drone fly is chunky and looks incredibly like a male honey bee. However, you will know that it is a drone fly because it only has one pair of wings, it has huge 'fly' eyes, antennae that you cannot see (you'd be able to see them on a bee) and its body does not have a narrow 'wasp' waist. Drone flies don't just mimic the appearance of honey bees, they even fly like them. There's no need for you to test this next fact, but it has no sting! This drone fly can be seen in gardens and hedgerows in almost every month of the year: the adults hibernate, but they can be seen on mild days in early spring.

Bee hoverfly

Volucella bombylans var. *plumata*

This hoverfly looks and sounds like one of the white-tailed bumblebees, but it has two wings, not four.

Large wasp hoverfly
Chrysotoxum cautum

This hoverfly looks very much like a wasp – the same shade of yellow – but it is slow flying and it hovers. It is a common hoverfly that you can find in gardens during summer months.

INSECT TONGUE LENGTH IS CRUCIAL

The length of a bee's tongue determines which flowers the bee can forage upon.

The reason plants have attractive flowers is to attract their pollinator, which is most often an insect. Once attracted, the flower provides an incentive for the insect to enter it, with the tiny drops of nectar held deep inside it. This is nature's way of ensuring that the bee has to reach far into the flower and become well covered in pollen on its way in, because the plant needs the bee to spread this pollen to other flowers of the same species. The bee has to have a tongue that's long enough to reach the nectar. In many flowers, only bumblebees and some moths have tongues the right length to do this, so only these insects can pollinate these particular flowers. By limiting the numbers of species that can reach its nectar, the plant increases its chance of successful pollination, because this means the pollen is more likely to be taken by an insect to another flower of the same species.

Bees gain all their food from flowers, yet plants' blooms vary greatly in their size and structure, because evolution has worked its magic so that different bee species are adapted to feed on different flowers. A bee with a long tongue, like the garden bumblebee, can reach nectar that other bees cannot reach and is able to drink from plants such as honeysuckle and *Buddleia*, which both have very long flower tubes.

A good example of the relationship between tongue length and flower shape is provided by the flowers of red clover and white clover. White clover has a much shorter flower tube than red clover, therefore if you look, you will notice completely different insects foraging on these two species: there could be honey bees, buff-tailed and early bumbles on the white clover, while on the red clover you are likely to see the common carder and garden bumblebees.

Realising that you are seeing different species foraging on different flowers – and knowing the bees' favoured food plants in your patch – will help you greatly from now on.

It's not just stripey insects that are important pollinators, moths matter, too! As well as beetles, flies and butterflies. Many of these insects are also important for plant pollination, and all of them play their role in maintaining biodiversity and feeding other species in the food chain. All of the actions recommended in this book will help all of these species.

FIVE SIMPLE THINGS YOU CAN DO EVERY DAY . . .

Stripey insects are so integral to life on Earth that many of our everyday actions can have an effect upon them.

1. Eat to support stripey insects and biodiversity

Maximise the proportion of fruits, vegetables, nuts and legumes in your diet. This is important because 70 per cent of global deforestation is caused by the need to clear land to grow soya for animal food, yet this forest habitat is crucial for biodiversity, and bees are at the heart of all biodiversity. Prioritising plant-based foods is essential for our efforts to combat climate change, too. There are thousands of species of bees on Earth, many of them unknown to us and not named by science.

2. Buy organic food

This is an important reason to buy organic food: organic certification is the best way to be certain that your food was grown with minimum damage to insect pollinators. Organic produce is more expensive to buy than conventionally grown products, and many of us have to budget carefully to shop this way, so one way to approach this is to calculate roughly how much buying organic will add to your weekly groceries and equate that cost to something else that you buy. For example, which would you rather have: good organic produce supporting good, sustainable farming for one week, or one bottle of wine? If you still feel uncomfortable about paying more, try to meet some organic growers and see the efforts they make to protect the environment, while at the same time having to compete with conventional farming's economy of scale.

Remember, too, that the apparent low cost of conventionally grown food is not a low cost for the Earth – there is a price to cheap food that is borne by our stripey friends being killed by pesticides, by harm done to the soil, by animal suffering and by loss of biodiversity. Surely, it's worth paying a little extra to support good, regenerative farming?

Another way to ensure that your food is grown without harm to stripeys is to grow your own! You'll need to rely on stripeys elsewhere to have done their work to have seed available for you, and stripeys in your patch to pollinate your plants.

77

3. Eat honey

By eating honey you are supporting a population of honey bees somewhere on Earth. You are also supporting beekeepers – people who work hard and are often fighting to retain natural habitats. Get to know your local beekeeper or buy certified organic honey that has come from a tropical forest.

Some honey nowadays looks a bit suspicious. All of the effort that goes into producing a jar of honey involves certainly as much work for the beekeeper as producing a bottle of wine or baking a cake. Yet some honey appears on supermarket shelves at crazy low prices, below what can be the minimum production cost of any honey. Fake honey does exist, so it's better to select some of good provenance.

One question that is often asked is, is honey vegan? Honey is an acceptable part of a thoughtful and good vegan diet, because supporting honey production can help to support the retention of biodiversity. For example, Bees for Development has for many years supported the beekeepers working in African forests. Local communities in Africa gain useful income from beekeeping in these forests. If not for their work, the trees would be cleared, with the tremendous loss of all the birds and animals living there. The forest would be briefly replaced by agriculture, growing maize, soya or other crops for a few years until the land became unproductive.

Although machinery makes it possible to destroy forest in a few hours or days, primary forest habitat will take hundreds or thousands of years to be restored. We cannot afford to lose any more forests, and beekeeping is one of the absolute best ways for people to generate sustainable income from them. Buying this honey and beeswax helps to support these people, these bees and these forests.

4. Supply water for honey bees

Honey bees need water all year. Even during cold winter days they will fly out from their hive in temperatures as low as 4°C (40°F) to find water. You can help by providing a safe supply;

for example, a wide, shallow dish filled with pebbles or moss so that they can land safely and not risk drowning. Birds and small mammals will appreciate this, too. In hot, dry summers all wildlife needs this water.

Swimming pools are great hazards for insects as they usually do not have shallow sides. Rescue any struggling insects by gently lifting them out, leaving them in a sunny spot to dry out, and helping them to recover by offering them a drop of sugary water.

5. Feed stripeys

Our stripey friends depend on nectar and pollen from flowers, and they need a continuous abundance of this food.

Spring is a crucial time for stripeys. Honey bees are alive all through the winter, but their honey stores may be exhausted by spring. Queens of many stripey species hibernate over winter, then urgently need food when they emerge in spring, while the young of some solitary bees emerge in spring and also need food.

Adult bees need to feed a mixture of pollen and nectar to their larvae. Pollen contains the protein that a larva needs to grow from egg to adult bee, nectar is their source of carbohydrate – energy. Adult bees must have nectar to enable them to keep warm and fly, but they eat little pollen as their growing is complete.

Summer is peak season for stripeys, with maximum numbers of insects seeking to feed on flowers and needing an abundance of floral food.

Autumn is another crucial time for stripeys: honey bees need to store nectar to have enough honey for the winter ahead, while queens of the other stripey species need to build their body reserves ahead of their winter hibernation.

Therefore, different stripey friends need nectar sources throughout all seasons: especially crucial are the scarce early spring and year-end sources of nectar. Climate change means that we have more variable weather patterns now; mild spells in early spring wake hibernating bees, but if this is then followed by severely cold weather, this can badly affect bee populations. Help bees by making sure there is food and shelter available. The key is to provide a succession of nectar-rich plants from early spring until late autumn – beginning with snowdrops and ending with ivy. If you live in a temperate climate, that is the ideal for a bee-friendly garden.

You also need to know that different bees forage at different times of day. For native plants and their pollinating insects, the plant will probably produce maximum nectar to coincide with the time of day that its native pollinator will be foraging. In general, the larger bumblebees can forage earlier in the day, and later in cool evenings, than smaller honey bees and solitary bees can.

. . . AND FIVE EASY THINGS TO STOP DOING TODAY

1. Stop weeding . . .

. . . and learn to love dandelions. Flowering early
in the year, these are a fabulous source of early
nectar and pollen for bees.

2. Mow less and make a meadow

Stop mowing and let your lawn become a mini meadow. Leave patches of grass to grow long and add in wild flower plug plants of devil's-bit scabious, oxeye daisy and the common knapweed, *Centaurea nigra*.

3. Eliminate the word 'weed' from your vocabulary

It is just a plant growing in the wrong place! Does it flower? If yes, leave it to feed the pollinating insect it is hoping to attract.

4. Don't tidy up until spring

Don't cut back dead stems of perennial plants until spring: once you see new growth appearing, that's the best time to remove the old. This is what happens in nature, and these stems provide crucial overwintering space for insects. The architecture of the winter garden can be

magnificent: seed heads are interesting and
beautiful, especially when patterned by frost.

5.   Do not think of buying or using insecticides, herbicides or fungicides

These all kill all bees and other stripeys. DEAD.

Insecticides kill all stripeys directly, fungicides
kill bees by killing the yeasts that they need to
digest pollen, and herbicides kill stripeys by
killing their food plants.

By killing insects, these products reduce numbers
of birds and animals higher up the food chain
that depend on insects for food. How does this
work? Of course, we never witness it, but as birds
and animals find less food to feed their young,
gradually populations decline.

WHAT TO PLANT IN YOUR GARDEN, PATCH OR POTS

Every scrap of earth can support a few plants, which will in due season flower and quietly provide food for stripeys.

BEE—FRIENDLY FLOWERS (BULBS AND SEEDS)

First let's look at bulbs to plant, then which seeds to sow. Snowdrops and crocuses are the best bulbs for our stripey friends; highly selected daffodils and tulips tend to offer little or nothing to insects, despite being some of the most popular flowers people choose to grow. People see dandelions as weeds that spoil their lawns, but don't pull them up: learn to love them as they are insect havens in an expanse of green.

Snowdrops, *Galanthus* species

If you notice a snowdrop flower is shaking, look inside and you may see a forager honey bee finding her first source of nectar and pollen for the year ahead. Early in the year, this is an encouraging sign of spring. The best way to establish snowdrops in your garden, in roadside verges or hedgerow bottoms is to plant them 'in the green', i.e., to split established clumps after they have flowered. This is quicker and more successful than buying and planting bulbs, so ask any friendly neighbours who have good clumps if they can share some. (You're going to have plenty of stripey-favoured plants to repay their kindness later on.)

Crocus species

Crocus provide a wonderful early source of protein-packed pollen for queen bumblebees. When you see a crocus flower vibrating, peer inside and you may see a queen newly emerged from hibernation and seeking her first food since last summer. If you buy and plant cheap mixed crocus bulbs, they will supply a month of food for bumblebees. Plant them in grass – the leaves soon die back after flowering.

Onions, garlic and all their *Allium* relatives

If you have onions or garlic in the kitchen that start to sprout green leaves, never throw them away! Plant them somewhere – any odd corner, at most times of year – and they will flower. The white flowers will be visited for a week or two by honey bees, bumblebees and hoverflies. You can also buy cultivated *Allium* bulbs; their huge purple or white flowers will be constantly visited by stripey insects during the 2–3 weeks that they flower. Chives, also a type of *Allium*, will be visited by red-tailed bumblebees and honey bees.

Bluebell

Hyacinthoides non-scripta

Useful because they flower early in the year; you will see honey and bumblebees as well as hoverflies foraging on them. If you see a bee carrying pollen out of bluebells, look carefully (or take a pic) and you will see that the pollen is dark blue-green, a very unusual colour for pollen.

Dandelion

Taraxacum officinale

This is a fabulous, important source of early nectar and pollen for the many stripey insect species that like dandelions, as the yellow flowers provide a stable platform for the insect to feed on the thousands of tiny flowers that make up one dandelion blossom.

It is possible to buy dandelion seeds, but most people will not need to, as this wild flower seeds with abandon. Dandelions tend to spring up in many places, so if you decide where you want to have the dandelions in the long term, deadhead the flowers before they set seed and the plant will live on, feeding insects for many years.

White dead-nettle
Lamium album

This common wild plant is an incredibly important food source for emerging buff-tailed and white-tailed bumblebees. There are cultivated 'editions' of white dead-nettle, but the native one is best.

Welsh poppy
Meconopsis cambrica

Once established, this poppy will self-seed and
go on providing forage for bees, year after year.
Once you have these in your garden, collect their
seed to share with friends. Every flowering plant
helps support our stripeys!

Foxglove
Digitalis purpurea

These are wild plants that also make wonderful
garden plants. Once established, they self-seed,
don't take up much space and will grow in slight
shade. Their large 'thimble' flowers are most
often visited by that longest-tongued bumble,
the garden bumblebee.

There are many beautiful cultivated forms of
foxglove, too: white foxgloves are among the most
elegant of flowers, and all seem to be well visited
by garden bumblebees.

Echium species

There are many species of *Echium* that provide wonderful food for bees, and chief among them is *Echium vulgare*, viper's bugloss (pronounced 'boo-gloss', not the unhelpful-sounding 'bug loss'!). Bugloss refers to the plant's leaves that are shaped like an ox's tongue – from the Greek *bou* for ox, and the Latin *glosso* for tongue.

Viper's bugloss is one of the best plants for bumblebees, in particular the white-tailed bumblebee, although it is visited by many types of insects, including painted lady butterflies.

Echium pininana is a 2m (6½ft) tall feast for bumblebees – incredibly special!

Phacelia tanacetifolia

Phacelia is one of the cheapest seeds to buy; it is used for green manure – which is when a plant is grown specifically to dig into the soil after it has grown, to improve soil structure

93

and fertility. It is a quick-growing hardy annual that can be sown almost any time of year, except in the middle of winter, and it will germinate rapidly and flower within 6–8 weeks, for a month or longer. *Phacelia* has blue flowers, dense foliage that smothers weeds, and an extensive root system that improves soil structure.

While flowering, it is continuously foraged by honey bees, bumblebees and hoverflies.

Hollyhock
Alcea rosea

Another of those plants that once you have it, you will never be without it, as it self-seeds with abandon. The important point is to have single-flowered varieties, not ones that have been 'improved' by breeding to display multiple frilly petals. Those 'improved' plants are of no value to any species except our own.

Mallow

Malva sylvestris

Another plant that's very easily grown from seed, and you will only ever need to buy one packet, as it self-seeds abundantly. Bees and hoverflies of many species love this plant, and often depart from the flowers dusted in its creamy coloured pollen. After it has finished flowering, you can collect the disc-shaped seeds and plant them where you want them to appear next spring. There are a number of different species in the mallow family – including tree mallow, which is a large shrub with grey-green leaves. All of them have similar flowers of great value to bees.

Marshmallow was originally made in Europe from the sap of the roots of mallow grown in marshes.

Penstemon

All varieties of *Penstemon* are visited by bees. It is easily grown from seed, and once established it will flower for a long period. The plants are perennials, too, so they will persist for many years, making them excellent value from one seed packet!

Cosmos

The open-flowered *Cosmos* will have stripey visitors all the time it flowers, between June and the first frosts. If you are going to grow this plant, you might as well go for the original large ones, those that reach at least 1m (3ft) tall, with each plant providing hundreds of flowers during the season. You must deadhead the flowers to keep them going all summer. Just one packet of seeds gives tremendous value for pollinators. Honey bees and buff-tailed bumblebees are particularly frequent visitors to *Cosmos*.

Poached egg plant
Limnanthes douglasii

Scatter this seed on open ground at the base of other plants where it can germinate in abundance and it will feed hoverflies and honey bees over a long period. Another self-seeder, once established it will come back year after year.

Sunflowers

Helianthus annuus

A cheap way to grow a lot of lovely sunflowers is to plant sunflower seeds intended for bird food – they geminate well and all are the 'unimproved' type that insects need. These are wonderful plants for bees, and after all that good pollination they will produce plenty more seeds to feed the birds.

Marjoram

Origanum laevigatum

This herb is loved by bees who work on it all summer long. A wonderful variety is 'Herrenhausen', which has dense pink flowers, is extremely easy to grow and self-seeds freely.

Rosebay willowherb

Chamaenerion angustifolium (Epilobium angustifolium)

Many gardeners regard this plant as a w**d (sorry, I cannot even write that word), but it is a wonderful plant and the 2m (6½ft) flower stems bring foraging insects to human eye level. Expect to see honey bees, white-tailed bumblebees and elephant hawk moths at night. Indeed, if you have just one or two of these beautiful flowers in your garden, friends may ask you what the amazing flower is, not recognising it as the pink flower familiarly seen *en masse* on railway sidings and waste ground. Of course, you will have hundreds more next year.

Sainfoin

Sainfoin is a traditional animal forage crop that's now re-emerging as a useful forage legume in agriculture and, like *Phacelia*, the seed is not expensive to buy. It is a perennial and flowers all summer long, providing copious amounts of pollen and nectar to honey bees, bumbles and solitary bees.

SHRUBS AND PLANTS

Common perennials and shrubs are often readily
shared by friendly gardeners and can usually be
found for sale in garden centres or supermarkets.

A NOTE TO BEE WISE WHEN BUYING PLANTS AND BULBS

If you are buying plants in a nursery, garden
centre or outside a supermarket or DIY
store, try to buy plants that are in flower and
that are definitely being visited by insects.
This is because horticulturalists have bred
some plants to make their foliage or flowers
more attractive to humans, and along the
way their ability to produce nectar and
pollen has been lost. Nobody notices or
suffers except insects. Of course, this test
will only work on days when bees are flying!

Many plants and bulbs are now labelled as being
'bee-friendly' or 'pollinator-friendly', usually

illustrated with a cartoon bee. This means that this plant species appears within a list of plants known to be good for pollinators. However, it does not mean that this plant you are about to buy is *safe* for pollinators – that is something quite different. It could well have been grown with systemic pesticide, which means the pesticide is inside the plant and can actually kill insects that feed upon it. Researchers at Sussex University have shown that the leaves, pollen and nectar of many of these plants contain a cocktail of pesticides.

If you are buying plants to encourage wildlife you don't want to accidentally poison them with pesticides.

We hope that the horticulture sector will soon realise that people like us want to buy plants that we know are safe for wildlife and do not contain pesticides.

Meanwhile, I suggest that you ask the seller of the plants what they know about their provenance: some, but not all, chains have

promised to not use systemic pesticides. It is
safest to buy from a nursery or local growers
who know the cultivation history of their plants,
preferably from an organic nursery, and watch
out for local plant sales on behalf of charities, etc.
These are the safest places to buy plants.

Ultimately, the best way to know that a plant
has been grown without pesticide is to grow it
yourself from seed.

Lungwort
Pulmonaria

This modest, small perennial plant supplies
precious early food for honey bees and
bumblebees. It is shade-tolerant. An interesting
aspect of *Pulmonaria* is that each flower lasts for
about a week, and during this period it changes
in colour from pink to blue. The bee you are most
likely to see visiting *Pulmonaria* is the hairy-footed
flower bee, *Anthophora plumipes*, which you
will see foraging exclusively on the pink flowers,
perhaps these are less likely to have been visited

already and contain most nectar. The male hairy-footed flower bees are ginger, while the females are all black with yellow hairs on the hind leg. After good pollination by these bees, many new plants will set seed, and you will soon have excellent abundance of this useful bee plant.

Spring heathers

Bees appearing from their winter hibernation need food, and flowers offering pollen and nectar are especially valuable to them at this time. Spring-flowering heather is one such plant, and it's particularly useful for window boxes and planters during winter. Heathers that flower during summer are also helpful.

Perennial wallflower
Erysimum

This amazing plant begins flowering early and continues for four or five months, supplying food for insects – the first visitors are early bees, joined by butterflies and hoverflies as the season progresses. Great for window boxes and planters. It is very easy to grow by taking a short cutting during spring and summer; select a non-flowering stem, cut it or just break it off 7–10cm (3–4in), just below a leaf joint, strip off the lower leaves and insert the cutting at the side of a pot containing a mix of gravel and compost. If you have a large pot you could make half a dozen cuttings this way, but don't let the leaves touch. Water, allow to drain, then place the pot in a plastic bag and put it somewhere warm and light, though out of direct sunlight. The cuttings will root within a month or so and can be moved on to another pot or planted out.

Aquilegia

These flower for a month, during which you will hear the garden bumblebee foraging noisily within their flowers, and they self-seed and multiply with abandon. Our stripey friends need the wild form or those with simple flowers like 'Texas Yellow', 'Florida', 'Colorado', 'Bunting' or 'Chaffinch'. There are many 'improved' varieties

with double flowers that insects cannot enter –
one example is 'Nora Barlow' – so be like stripeys
and ignore those ones.

Scabious

Field scabious *Knautia arvensis* is a wild flower,
nowadays seen occasionally on road verges,
which is attractive to many insects. These and the
related *Knautia macedonica* are widely available
in garden centres, and these small perennials
will keep flowering over summer and autumn as
long as you remove the dead flower heads. The
flat faces of the flowers provide excellent feeding
platforms for bees, butterflies and hoverflies.

Comfrey
Symphytum species

There are many species of comfrey and all are
popular with our stripey friends. In early spring,
a bed of comfrey will be quivering as queen
bumblebees feed on its welcome nectar.

Thistles

Asteraceae family

Stripey insects forage on flowers of all members of the thistle family, from the gigantic flowers of artichokes, popular with red-tailed bumblebees, through the globe thistles loved by honey bees, to the far smaller flowers of wild thistle foraged by honey bees, bumble and hoverflies.

If you grow *Echinops ritro*, the globe thistle, in your garden, you will be rewarded three times – by the beautiful spherical blue or white blooms, by the fact that while in flower they are rarely without honey bees or bumbles, and finally, in autumn, you may see goldfinches feeding on the seed heads.

Elecampane

Inula helenium or *Inula hookeri*

Inula are large and robust plants with yellow flowers in late summer, on which you will see a wonderful variety of bees, wasps, hoverflies and butterflies. All species of *Inula* spread well, and

if you have a friend with some in their garden,
ask them if they don't mind sharing some roots.

Russian sage
Perovskia atriplicifolia/Salvia yangii

This long-blooming perennial resembles lavender
and will be visited by many insects while it flowers
continuously during summer. Towards the end of
the season, stop deadheading and the seeds will
feed finches and sparrows. Next spring you can
cut back what is left of the dead flower stems.

Lavender
Lavandula species

Lavender is a Mediterranean plant that thrives in
full sun, whether in a pot, window box or garden.
A hedge made from lavender is a wonderful sight in
summer, particularly with an abundance of foraging
insects adding a whole extra dimension of interest.

Taking lavender cuttings during summer is quick
and easy: they root readily and you can have as

many new plants as you want – all for free. Just select some side shoots 10–15cm (4–6in) long from the main plant and, one at a time, pull them away from the main stem with a thin strip of bark still attached. Remove the lower pairs of leaves so that the cutting now has about 5cm (2in) of bare stem, then push this stem down into a small pot of gritty, peat-free compost near the edge, ensuring that the cutting is held firmly in place by the compost. You can place about ten cuttings around the sides of an 11cm (4in) pot to give you ten new lavender plants in 6–8 weeks. Water well and leave in a warm, shaded place. After about a month, rooting will have started, then you can pot them up individually when they are beginning to fill the original pot.

A NOTE ON 'MODERN' LAVENDER

Bees do not visit some 'modern' lavenders –
those that have been selected by humans for
their looks rather than by nature for their nectar.
Therefore, make sure that you have a lavender
that is popular with bees by taking cuttings from
a plant that you can see is being well visited by
stripey insects. If you are buying lavender, take
notice if it is already being visited by insects.

Cotoneaster

Cotoneasters are widely available and some of
the cheapest shrubs to buy, existing in every
size from dwarf plants to large hedging types.
Cotoneasters have flowers that are smothered
with honey bees in summer and provide valuable
berries for birds in winter.

Cotoneasters are easy to propagate by cuttings,
as described for lavender, although they will take
longer to root and grow on. Another easy way to
gain more cotoneaster plants is by layering – this
works well for the spreading, low-growing types

of cotoneaster. Just cover a thin branch near the ground with soil and firm it down. This length of branch will now form roots into the ground, and when well developed, the new plant can be severed from the original and planted elsewhere.

Cotoneaster horizontalis is so easy to cultivate that it is now considered an invasive pest in some areas of the UK and it is illegal to plant this species in the wild.

Hybrid germander
Teucrium x lucidrys

Hybrid germander is like thyme but larger, more robust and with larger flowers. It blossoms over a long period in midsummer and is absolutely loved by honey bees, bumbles and butterflies. It can even be used to create a hedge: a good alternative to replace box if your plants are suffering from box blight.

Rose

Most cultivated roses are of no value to insects, as double flowers with multiple petals mean that bees

cannot reach any nectar or pollen stored within.
The roses that insects need are those with simple,
open flowers, where a visiting insect can access the
nectar and pollen. Examples are all the 'wild' rose
species like *Rosa arvensis*, *Rosa moschata*, *Rosa
moyesii*, *Rosa mundi*, *Rosa gallica* and *Rosa rugosa*,
as well as roses that have not been 'improved' too
much and still have open flowers like 'Ballerina',
'Buttercup', 'Iceberg', 'Kiftsgate', 'Morning
Mist', 'Paul's Himalayan Musk', 'Sally Holmes',
'Scarborough Fair', 'The Lady of the Lake', 'The Lark
Ascending', 'Rambling Rector' and 'Wedding Day'.

Red bistort
Persicaria amplexicaulis

There are several *Persicaria* species commercially
available, and not all are excellent for feeding
insects. However, *Persicaria amplexicaulis*
'Firetail' is a wonderful bee plant, flowering
continuously during summer and constantly
visited by honey bees, and later on by wasps
looking for the last of the summer's nectar.

Salvia species

There are hundreds of different salvias, including common sage, to choose from and all except the red bedding types will be visited by a good range of stripey insects. *Salvia* species include annuals, hardy herbaceous varieties that come back every year, and woody types, like sage itself, that persist over winter. Being Mediterranean plants, they all love full sun. *Salvia* 'Amistad' and *Salvia* x *jamensis* 'Nachtvlinder' are wonderful plants for bees, flowering all summer and well into autumn.

AVOID BUYING USELESS PLANTS

Please remember plants produce flowers to attract insects, not to please humans. Yet human titillation, and profit, are all that many modern 'improved' plants achieve during their brief lives. Try to avoid buying attractive but useless bedding plants like busy lizzie, double begonias, French marigolds, pansies, petunias and pompom dahlias.

TREES

The best trees to plant in any situation are native tree species, because they will support native insect populations and thereby everything else higher up the food chain. All nine trees listed here are native to Europe. Different species of the same genera exist in the Americas and in Australia; *Eucalyptus* and *Callistemon* species are wonderful sources of nectar and pollen for insects.

One flowering tree has thousands and perhaps millions of flowers, so it offers far greater value for stripey insects than could be provided by flowering plants covering the same area of ground. Therefore, if at all possible, plant trees to greatly add to the available forage for our stripey friends, and to enhance the environment for everyone.

Willow *Salix* species

The many different species of willow tree range from small trees to huge ones, and all provide useful early pollen sources; as our winters become warmer, these early food sources are becoming increasingly important for stripey insects.

When a colony of honey bees or bumblebees gets going in early spring, the queen is daily laying many eggs and there is a need for pollen to feed the larvae as they hatch. It is usually the availability of fresh pollen that is the critical factor, and what makes early flowering plants like willow so useful to bees.

Willows are one of the easiest trees to cultivate; just take cuttings from an existing tree by cutting a healthy piece of branch about the diameter of a pencil and 30cm (12in) long, push it into the ground sometime in late autumn or early spring, keep watered in dry weather, and stand back to watch it grow.

Ornamental willows vary in their usefulness to insects, so if you buy a pussy willow plant from a garden centre in spring, look for plants already being visited by stripey insects.

Blackthorn or sloe
Prunus spinosa

Blackthorn blossom is often the first to be seen each year, with its abundance of white flowers carried on black stems. It supplies early nectar for honey bees and emerging bumblebee queens. After its fabulous early spring display, the bush takes a low profile while its fruit – the sloes – mature. The density of a blackthorn bush makes it excellent for providing safe refuge and shelter for birds and other wildlife.

The sloes are like small, intensively sour plums with a blue bloom, which gradually darken to glossy black. Tradition has it that they should not be picked until after the first frost of winter, and then used to make sloe gin. Of course, the birds may reach them faster than the sloe-gin maker!

Cherry

Prunus avium

In early spring, woodlands full of leafless trees still appear dark, but a first dusting of blossom is provided by early flowering wild cherry trees: a welcome nectar source to early flying stripey insects.

Ornamental cherry trees can be good for bees as long as the flowers are still the single type – not bred to have double flowers that insects cannot penetrate.

Apple

Malus domestica

All apple cultivars, whether ornamental crab apples or for edible fruit production, supply valuable nectar for bees and, indeed, apples need pollination to produce the best harvest of top-quality fruit. Even if the fruit are not picked and lie on the ground below, they provide much-needed food for migrating flocks of thrushes in autumn and winter.

Hawthorn

Crataegus monogyna

Hawthorn makes up the main tree of many hedgerows and is an extremely valuable species, providing important safe refuge for wildlife, but its flowers also provide nectar and pollen for insects, while the autumn berries, and the visiting insects, too, provide food for birds. Bare-rooted hawthorn plants can be bought very cheaply from late autumn onwards and are best planted during winter months.

Mechanical cutting of hedgerows at the end of summer removes the season's fruits, seeds and berries that sustain wildlife through autumn, and flowering tips cut now will not flower next spring. So hawthorn hedgerows should be trimmed – not annually – but with a (maximum) three-year rotation. It's time to encourage people responsible for hedgerows to manage them with a care for wildlife (see Begin stripey awareness in your community on page 148).

Mountain ash, or rowan

Sorbus aucuparia

A small tree that will grow in many difficult spots, including polluted and exposed areas. The white flowers in spring are much visited by stripey insects, and the flowers are followed by attractive, orange-red berries that hang in bunches. Blackbirds and other thrushes welcome these berries during autumn and winter.

Horse chestnut

Aesculus hippocastanum

These are wonderful trees for bees, their pink or white 'candles' of flowers in late spring provide abundant nectar and pollen for many types of insects, including bees.

Horse chestnut pollen is brick-red and when the chestnuts are in flower, beekeepers see bees carrying loads of this unusual dark-red pollen into the hives.

The cheapest way to obtain a chestnut tree
is by planting a conker! Gather fresh conkers
in autumn, put them in the fridge for a while
(they need a period of winter cold to trigger
germination), then plant them. The young
chestnut will grow rapidly and begin to flower
after 4 or 5 years. Do not plant in the wrong
place, though, as the tree will eventually become
a tall (30m/100ft) and wide tree!

Lime
Tilia species

During early summer evenings, the air near to
lime trees is filled with their delicious sweet
scent. And the lime tree itself will seem to be
humming, with many stripey friends feeding on
the nectar of the lime.

Please know that this lime is nothing to do with
the citrus lime that is so delicious in a mojito!
The tree lime for bees is the European *Tilia*,
known as lime in English and linden in European
languages, and it plays an important role in

European culture and history, being considered a mystical tree of life that symbolises health, justice, victory and fertility. The avenues of flowering lime trees often planted in European towns and cities are of tremendous value to urban stripeys. (See page 141 for how to make lime flower tea.)

Spanish chestnut

Castanea sativa

In late spring, the Spanish chestnut has beautiful flowers, and these are full of bees gathering nectar and pollen. Honey bees create a dark and bitter honey from chestnut, which is much sought after and absolutely delicious.

HOMES AND LODGINGS

As well as food, stripey insects need safe places to live. Our increasingly concrete and tidy landscape removes those possibilities, and you can easily help to replace them by deliberately leaving alone the untidy corners that nature needs. Most bee species spend the greater part of their life as egg, larva and then pupa, and it is only in their final, often brief, adult stage that we see them on the

wing. This is why a safe nesting habitat
is crucial.

Here are five types of bee housing and lodging,
so you can find something for your garden.

THE SIMPLEST BEE HOMES FOR SOLITARY BEES

As you might guess from their name, carpenter bees make their own homes by tunnelling into dead wood, and many other bees, beetles and other tiny wildlife will appreciate the useful, safe habitat provided by an undisturbed heap of sticks and logs. As the pile gradually reduces in height over the years, keep adding on top. And as mentioned below, hollow stems left over winter will be used by many overwintering stripeys.

Areas of bare earth and 'mini cliffs'

Most species of solitary bees are mining bees, which cannot use the bee hotels with tubes described below. These bees need to nest in open soil, banks or short grass, so a good wildlife garden should offer this type of habitat too.

A five-star bee hotel

The idea of the bee hotel is to provide a safe place in which solitary bees can nest over winter. In nature, solitary bees nest in hollow stems and the tiny nooks and crannies that occur in a natural environment; however, in our increasingly manicured, trimmed, tidied, strimmed, cut and concreted-over world, these untidy corners are increasingly hard for bees to find.

You need to have your hotel ready for your guests by early spring. It's easy to make: first you need a container. It must be at least 20cm (8in) deep, have an open front, three walls, a floor and a roof, with a 'porch' to protect the contents from rain. It can be any container you have to hand – maybe a wooden box or drawer: even a cardboard Tetra Pak container could do, as long as you protect the contents from winter weather. Or if you are handy with tools, it's a simple bit of carpentry: cut a plank of wood to make a four-sided box, with a longer piece for the back to allow for attachment to a fence or wall, and a longer piece for the top to overhang the cavity. Or you can use a piece of drain pipe as a container, though you must ensure you add a back to it – maybe by placing it flush against a wall or fence.

Now fill your container with tubes. These could be paper straws, but more substantial would be bamboo canes cut to length, hollow plant stems you have collected, or cardboard tubes (you can buy special tubes for bee nesting). If you are using bamboo canes, cut them just below a node (a bulge in the cane stem), as this will provide a sealed end for the tube. If you use a stem with a pithy centre, the bees will dig out the pith to make a cavity for their nest. Suitable plants with pithy stems are *Angelica*, elderberry, goldenrod, hydrangea, *Leycesteria*, raspberry and sunflowers, among many others. The tube length should be roughly 15cm (6in), and it is good to have some variation in length, so that at the front of the bee hotel the tubes are not flush with one another and each bee can identify their own hotel room. We've all faced that problem!

Now you need to situate your hotel somewhere attractive for your guests: it must be firmly attached or lodged, at least 1m (3ft) up, ideally facing south or west, but not into prevailing wind, and of course protected from rain or snow. Direct sun could also make the tubes become too hot, and the eggs will be killed if the nests become wet. Experiment over a few seasons to find the best place for your local bees and weather.

Your likely guests will be the red mason bee *Osmia bicornis,* which is active in spring and early summer and needs a hole diameter of 7–8mm, and leaf-cutter bees, which are active in summer. A range of tube diameters between 4 and 10mm will attract a range of solitary bee species to your hotel.

If the eggs remain safe through winter, next year the larvae will hatch, survive on the food left for them by their late mother, and emerge as adult bees, to live for a month or so, pollinate their food plants and prepare for their next generation of bees.

In nature, bees would not nest in the same tube, year after year, and nor should they in your hotel. It is important to clean it from one year to another and replace the tubes to prevent any possible accumulation of diseases or parasites, which become more likely when the bees are being encouraged to nest closely together in this way. You must attend to this: you don't want your bee hotel to become a bee hospital!

And, of course, the best and easiest way to provide bee hotels is to resist being a fanatical garden-tidier. If you leave plant stalks intact over winter and do not cut them down until late spring, then they will be natural hotels to many overwintering bees.

HOMES FOR BUMBLEBEES

In nature, bumblebees need somewhere safe and dry to nest – that could be a hole in the ground, a disused bird or mouse nest, or the base of tangled vegetation in grassland. They often nest in small cavities under garden sheds and compost heaps. Their nest, consisting of wax cups holding honey, stored pollen and the developing young bees, is no larger than a grapefruit. Bumblebees very rarely sting – although perhaps if you trod on one in your bare feet, it might sting you.

Like everything else trying to survive in a human-dominated world, bumblebees also need safe places to live as well as nest, and you can help to provide this. The easiest way is to put up more bird boxes – if they are not occupied by birds, they are likely to be inhabited by tree bumbles (see page 42). To accommodate other bumble species, maintain a garden with plenty of undisturbed nooks and crannies for bumblebees to nest inside – I find that upturned clay flowerpots set on rough ground so that the bees can access them, with some nice moss inside and a slate to cover the flowerpot's hole, are readily adopted as home by bumblebees. A more professional way to provide this is to situate a piece of hose pipe to serve as an entrance and submerge the rim of the pot and the pipe below ground level, to make the whole thing more stable. A few holes punched into the pipe are important to remove any water that might accumulate there and adding some wire gauze to lift the nesting materials above the earth will also serve to keep them dry and make everything cosier. Bumblebee nests like this are an ideal safe home, are quick and cheap to make and can be left in place for many years.

Do not be tempted to buy the bumblebee colonies offered for sale, as these bees are reared intensively, are unlikely to be your local race of bee, and provide multiple opportunities to further stress the local wild bumblebee population by introducing new diseases and associated viruses.

PROVIDE A HOTEL FOR HONEY BEES

In nature, a honey bee colony would choose to build its nest inside a tree or other natural cavity. Yet nowadays there are increasingly few old, hollow trees, and wild honey bee colonies tend therefore to build their nests inside building cavities and chimneys and are too often perceived by humans as a 'nuisance'. A honey bee colony needs room to accommodate its large colony size and honey store for surviving winter – this requires a cavity of 40–50 litres (9–11 gallons). Should you wish to obtain one, Bees for Development provide a Bee House with a cavity of this size, which serves as an artificial hollow tree and simply gives a safe nesting space for honey bees.

World-renowned scientist Professor Tom Seeley has endorsed this Bee House design:

'Throughout the honey bee's vast range of Europe, western Asia and Africa, it lives both as managed colonies in man-made hives and as wild colonies in natural cavities. The Bee House sold by Bees for Development matches the housing preferences of European honey bees, so by mounting one on a tree or building, you will help sustain the population of wild honey bee colonies in your region.'

Professor Thomas D. Seeley, Horace White Professor in Biology, Department of Neurobiology and Behavior, Cornell University, USA

Inside the box is a cavity of just the right size for a honey bee colony to build its nest. The space has a flat ceiling, with a loft space insulated with wood shavings. The ceiling is where the honey bees will attach their combs, which are built from beeswax and are where the honey bees will live and rear their brood (young bees) and store their food (nectar being converted into honey and pollen). The bees' only entrance to the box is by the floor at the front.

During the honey bee swarming season, in late spring and early summer, if there are existing honey bee colonies in your area, scout honey bees will begin examining the box entrance and going inside to 'measure up'. If the scout bees judge the box to be a good nesting place, a swarm of honey bees may arrive in the next few days or weeks.

The Bees for Development Bee House is already rubbed inside with beeswax to create a 'bee-friendly' aroma. Bees may be further encouraged to settle in the Bee House by rubbing around the entrance with beeswax, or with the tiniest drop of anise extract (from the seeds of *Pimpinella anisum*), or other herbal oil to attract the scout bees.

This Bee House simply provides a nesting place for a honey bee colony. If you want to harvest honey from bees then you need to keep them in a different type of beehive. You cannot open the Bee House: the bees will build their nest attached to the ceiling (and maybe the walls) of the box and will seal up all cracks and gaps with propolis (resins that they collect from plants) and will keep it clean inside.

Like any other animal, bees nesting in the Bee House may succumb to disease. The introduced exotic mite, *Varroa destructor*, will inevitably be present in the colony. However, our experience is that wild-nesting honey bees survive well in the presence of the mite. If a swarm occupies the Bee House, and if the bees are susceptible to *Varroa*, the colony is likely to die at the end of its second year.

By helping to increase the population of honey bees, you will be helping to restore and maintain the genetic diversity of honey bee populations. Honey bees are evolving to survive in the presence of *Varroa*, and the greater the population of honey bees, the greater the genetic pool to develop resistance to the mite.

In addition to the *Varroa* mite, other predators include ants and wasps. A strong honey bee colony will defend itself well from these predators, which are part of nature. Other exotic predators may be introduced in the years ahead. The best way for honey bees to survive in the presence of these predators is by maintaining a large and genetically fit population capable of evolving and allowing natural evolution – survival of the fittest – to take place. By providing homes for honey bees, we are helping with this process. In nature, bees nest in cavities in trees and some people believe that they benefit from the microflora present in trees.

If the honey bee colony becomes weak or dies out, wax moths will enter the box and consume all the comb and remnants of the bees' nest. Eventually, with no food left, there will be no wax moths left either. At this stage, the box now has a lovely propolis (plant resin) lining and all the good honey bee aromas, making it ideal to be re-occupied by another honey bee colony as soon as the next swarming season arrives.

The empty box weighs 10kg (22lb), but when occupied by a honey bee colony it weighs up to 30kg (66lb). Therefore, it must be situated in a very safe and secure place. The box should be positioned 3–5m (10–16ft) above ground level, which will make it more attractive for honey bees to occupy. Ideally, it should be sheltered from strong winds and shaded from too much sun. We have found it works well in a tree, on a flat roof, on a balcony or attached with strong fixings to the wall of a building.

Of course, you can make your own honey bee house – the important factor is to ensure a cavity space of at least 40 litres (9 gallons).

FUN PROJECTS FOR EVERYONE

At this crisis stage for the Earth's biodiversity, everyone can help to make a difference. There are feasible, low-cost things that you can do, and your actions will encourage others to help too.

Create a mini meadow

If you have a lawn, make this the year you begin to see it for what it is: a monoculture of no value for life on Earth. Change your mowing regime and begin turning it into a meadow teeming with a range of wild flowers and insect life. Once established, your meadow will be greatly more useful for stripeys, more interesting for you, be little work – and all with no costs involved.

'Improved grassland' is what you mostly see covering farmland intended for grazing animals, and golf courses and lawns all aim to be a uniformly vibrant green and weed-free. It might as well be green concrete as far as insects and wildlife are concerned. These lush lawns consist mainly of ryegrass and other selected grass species, fed with nitrogen and phosphorus fertilisers to keep the vigorous grasses perennially bright green, although sometimes a few flowers like buttercups, docks and thistles can survive the regime.

The nitrogen and phosphorus from artificial fertilisers are persistent in soils, and these nutrient levels have to be reduced to return these highly fertile sites to flower-rich areas: a meadow teeming with wildlife requires natural, lower levels of nutrients to permit a wide range of species to survive together.

If your lawn has been seeded with rye grass you will need to take this off by completely removing the turf layer. This gives you the chance to sow seed of local wild plants on the poorer subsoil. Common in such seed mixtures are oxeye daisies, favoured by hoverflies.

Natural meadows in Europe contain a plant named hay rattle. This yellow flower has papery seed capsules within which the ripe seed rattles when mature in summer – the rattling of the seed was an indication that the meadow was ready to be cut for hay, hence the name 'hay rattle'. In fact, hay rattle is an especially useful plant for restoring meadows, as it helps to reduce soil fertility: it parasitises the roots of grasses enabling other species to thrive.

You can use plug plants to rapidly gain the flora that you want – this is quicker but more expensive than using seed, or simply wait for nature to take over. The quickest (and probably most expensive) way to gain a meadow is to remove the turf and replace it with meadow matting – matting impregnated with wild flower seed.

139

Do not mow it this summer and see what you have: you may be surprised and delighted to see the beautiful grasses flowering in midsummer. You maybe will find flowers of clover, vetch and hawksbeard. If you fear the wrath of neighbours, carefully mow a path through the developing meadow, as this relays the message that you are not being lazy, you have just implemented a new no-mow management regime. Remember that unless you are still pushing and pulling a rotary mower, mowing a lawn burns fossil fuel and destroys biodiversity at the same time.

Cut your developing meadow only after everything has finished flowering – this will be in late summer. Allow the cuttings to lie on top of the lawn for a few weeks so that seed can fall into the soil, then remove them – this is important to gradually reduce the fertility of the area.

Once you begin to establish a range of plant species in your meadow you will find it far more interesting than boring green grass, although it is never going to look so unnaturally manicured as a grass lawn.

Learn to make bee teas

I'm talking about making tea from some of the
bee plants you are growing! The essence of
nature is to produce in abundance, supporting
many species – including us. If you pick your
flowers away from roads or industry, and where
nobody is using agrochemicals, your teas will
be as good and safe as possible.

Many of our stripey insects' favoured plants can
be brewed into tasty teas, and most of them
are believed to have medicinal or therapeutic
properties. To extract the best of the marvellous
aromas and compounds, mix a handful of the
fresh leaves or flowers with boiled water. You will
have to experiment to get the strength that you
like, by deciding how long to leave it to infuse.
Some need steeping overnight to achieve a really
intense brew.

Make lime flower tea in early summer by picking
a handful of fresh flowers, infusing them in
boiling water, then straining them off. You can
dry the flowers, too – leave them on a clean tea

towel or piece of paper and when completely dry, store in an airtight container to use throughout the year.

Other bee plants from which you can make good teas are dandelion (leaves and root), all of the mints and sages, rosebay willowherb leaf, marigold flowers, fennel, thyme and many other herbs.

The one tea that you will not want to drink is 'comfrey tea', which is the name given to comfrey leaves left to break down in water over a few weeks. It makes a fabulous liquid feed for plants, but, being full of nitrogen and breaking down organic matter, it smells dreadful!

Become a rebel botanist

If you are becoming stressed by fellow citizens' disregard for nature or general 'plant blindness', help everyone to learn a little by chalking plant names beside interesting plants growing through cracks in pavements and striving to thrive in odd corners of the urban environment. Who would not like to know the lovely names of enchanter's nightshade, Good King Henry and forget-me-not? Knowing some vocabulary of nature is the first step towards appreciating it.

Or follow in the footsteps of early rebel botanist Ellen Willmott (1858–1934), who secretly planted seeds of the giant prickly thistle *Eryngium giganteum* wherever she went. Now known as 'Miss Willmott's ghost', this thistle is a wonderful bee plant. Deep inside her handbag, along with the *Eryngium* seeds, Miss Willmott also always carried a revolver – taking rebel botany further than I recommend!

There is great satisfaction to be gained from seeing seeds you have sown flourishing and being foraged by stripey friends, unnoticed in urban corners and supermarket car parks. Good seeds for some rebellious botany are dandelion, rosebay willowherb and thistles. A gentle word of warning, though – only do this on waste ground in urban places, **never** contaminate existing good natural habitat with introduced seed.

Create a mini forest

Follow the wonderful work of Akira Miyawaki, a botanist in Japan who has won the Blue Planet Prize for his idea, which is now taking hold worldwide. If your community has access to a space about the size of a tennis court you can plant a mini forest, with 30 or more species densely planted to recreate the layers of a natural forest. It will quickly create a very dense area, faster and with more biodiversity than is achieved by conventional planting.

Take a walk after sunset and see the elephant in the garden

Butterflies appeal to everyone, and it should be easy to see them in summer on plants like *Buddleia*. Butterflies belong to the Lepidoptera type of insects, yet they are only a tiny number of species compared with moths, which make up 95 per cent of Lepidoptera. There are thousands of species of moths in every country, yet they stay little known because they mostly fly at night.

So take a walk with a torch as night falls, and in summer months, if you know where rosebay willowherb *Epilobium angustifolium* is growing, look for the elephant hawk moth. In spite of their exotic beauty, these are not rare moths and you have a good chance of seeing one.

Fundraise for a charity

Do contact your chosen charity and ask them what help they need. If, like most charities, they need financial support, there are many ways you can help. One way, which costs nothing, is to spread news of their fundraising campaigns by social media – every charity will appreciate your help with this. You can also help bee charities' aims by explaining with simple signage why you have not mowed your grass and why you are not using any pesticide. This all helps to spread the word about our need to protect biodiversity, for if we get it right for insects, we get it right for everything.

People have raised money for my charity, Bees for Development, in all sorts of imaginative ways: Poets for the Planet organised a Poem-a-thon, people sew and sell anti-Covid masks made from bee material, and make soaps and candles made with beeswax to sell. Beekeepers may follow the example of HRH The Duchess of Cornwall, a keen beekeeper herself, who donates the proceeds from selling her honey to our charity.

Bees for Development is the longest-established charity working to support bees and beekeepers in the poorest nations: aiming to alleviate poverty and at the same time improve biodiversity. In recent years, the charity has become involved with bee conservation and training about bees and beekeeping in wealthier nations as well as less privileged ones. It is a small and growing charity with headquarters in Monmouth, in south Wales, UK. Please visit www.beesfordevelopment.org if you would like to know more.

Take some stripey portraits

Taking a quick snap with your phone can be a great way to give yourself time to figure out a stripey's identity. If you slowly move in close to get your pic, you will often find that a bumblebee waves at you by raising a leg to politely ask you to keep your distance.

Flowers with an open, flat structure, such as dandelions, elecampane or sunflowers provide a nice flat platform for stripeys that makes them easier to photograph. Find a plant that is being visited by stripeys and select one flower that is in perfect condition, well lit, and with a good background. Set up your phone or camera on a tripod, compose the pic, adjust settings and focus. Now sit back and wait for an insect to visit the flower and use continuous shooting to capture that perfect shot.

For the best pictures of stripeys, you will need a macro lens with a focal length of 100mm or more – this will enable you to achieve life-size

magnification at a practical working distance
away from the insect.

Begin stripey awareness in your community

In 2020, we established Monmouth as the UK's
first Bee Town. What does this mean in practice?
That people living here are well informed about
our need for stripey insects and are doing
all that we can to support them. It's not just
individuals, the County and Town Councils
have adopted Pollinator Policies, abandoning
the herbicide glyphosate, and are doing their
best to help. And schools, local groups focusing
on green, conservation and climate change
issues, clubs and associations are joining in too:
stripeys provide a good way to mobilise fellow
citizens to take an interest in saving bees and
restoring biodiversity and bioabundance. Stripeys
are proving themselves even more useful by
providing a popular topic around which we can
unite and engage more people and organisations
to become involved.

Write to your local politicians

The two main reasons for insect decline are:
1) habitat loss and 2) pesticides. These two
work together because habitat loss is due to the
intensification of farming and its associated use
of pesticides.

Every nation needs to set targets now to achieve
substantial reduction in pesticides. The EU
recently announced its aim to achieve 50 per cent
reduction in pesticide use in coming years, and
this is a good ambition for other nations to adopt.

Farming has to become more sustainable and
wildlife-friendly if we are to restore insect
populations to the levels of even 40–50 years ago.

Write to your local council

Local authorities tend to receive many complaints
from the public they serve, yet they rarely receive a
compliment. If your local authority makes an effort
to mow less or provides excellent forage for bees, let
them know that you have noticed and are delighted

by their good work. It is so rare for these bodies to receive a compliment that it will greatly encourage the people working in this department! All councils should have appropriate Pollinator Policies and should be implementing them. The herbicide glyphosate has been completely banned or restricted in 21 nations, and many towns and cities, such as San Francisco and Seattle, are taking this path too. Germany will have a complete ban in place by the end of 2023. In the UK, parks in North Somerset and councils in Bristol, Glastonbury, Lewes, Shaftesbury and Wadebridge have stopped the use of glyphosate and are trialling other options.

Bee Friendly Monmouthshire has prepared a Hedgerow Manifesto, setting out good practice for managing hedgerows – please encourage your local council to adopt similar practices. (Go to www.beefriendlymonmouthshire.org to download.)

Bee-friend a beekeeper

Beekeepers are always to some extent botanists too, and they will be able to tell you much about the local flora, being well informed by their bees through the colour of the pollen they bring back to the hive, and the honey they produce – or you could become a beekeeper yourself! Find a local course (see page 152) and beware, once you are interested in beekeeping, there is no going back – it's lifelong learning and you are joining a community that is thousands of years old. Bees are the best pets!

Attend a bee event

The world's best honey show takes place at the end of October, currently the venue is Sandown Park in Surrey, in the UK. While this event is mainly attended by the beekeeping world, any reader of this book would find much of interest, and you will see beautiful honeys of every hue lined up in competition – there is no honey show anywhere else to rival this one.

www.thenationalhoneyshow.co.uk

Celebrate World Bee Day

This takes place every 20 May – hold a bee party and raise funds for stripeys!

Attend a bee course

Wherever you live there will be a local beekeeping association offering courses for beginners, often with free 'taster' days. Or you could join one of Bees for Development's courses – there are weekend courses on sustainable beekeeping, on weaving skeps (baskets for bees), and many other topics – see www.beesfordevelopment.org.

Plant a hedge

A living hedge is of far greater beauty and value to nature than any wall or fence, and bare-rooted hedge plants can be bought inexpensively for planting during winter months. Please select plants that provide good forage for insects – for example, hawthorn for large hedges, lavender or hybrid germander (see page 109). If you do not have a garden, many conservation organisations would welcome your help with the important work of restoring hedgerows and other restoration work during winter.

Learn to love ivy

Have you noticed the unusual feature of common ivy *Hedera helix* – that as the plant becomes older it changes its leaf shape? The classic 'ivy-shape' leaves on long climbing stems belong only to young plants: once the plant is well established (maybe after ten years), it becomes an adult and the leaves become rounder, and the plant becomes shrubby and covered with flowers in autumn and berries that ripen over winter. It is in this adult state that ivy is so valuable for wildlife – look at the flowers on a sunny day in autumn and you will see it covered in late flying insects, and in particular the ivy bee – a mining bee species on the wing only towards the end of year when ivy is flowering. Ivy flowers are incredibly valuable for honey bee colonies too, as the very last nectar source before winter sets in.

Choose organic cotton

Organic cotton is becoming increasingly available, and it is a particularly good way to support regenerative farming, as conventional cotton farming uses colossal amounts of pesticides. Organic certification does not just relate to the use of pesticides, it applies to the whole of the cotton's cultivation and processing. Organic cotton farming methods create resilient crops by building healthy soils, using inputs such as natural fertilisers and sustainable water management. There are amazing environmental savings; for example, production of a simple cotton T-shirt weighing 150g (5oz) uses 704 litres (186 gallons) of water, compared with 8,207 litres (2,168 gallons) using conventional methods.

CONGRATULATIONS!

Well done indeed! If you have already planned or completed some of these ideas to make life better for our stripey friends, you can be confident that more of them will be alive this year and next because of your efforts. Your final project is to spread the word, so that more people understand the problem and what they, too, can do to help.

FURTHER INFORMATION

Gradually some useful phone apps are being
developed to help with identification – for
example, *Wild Bee ID* in North America, and
What's that Bumblebee in the UK.

Worldwide

The Bees of the World by Charles Michener
(John Hopkins University Press, Baltimore,
2007) – the definitive guide to 20,000 named
bees.

Australia

A Guide to the Native Bees of Australia by
Terry Houston (CSIRO Publishing, 2018) – a
wonderful book.

North America

The Bees in Your Backyard by Joseph S. Wilson and Olivia J. Messinger Carril (Princeton University Press, 2015) – a brilliant resource for stripey spotters in North America. It would be wonderful if we had guides as good as this for every world region.

UK

Bumblebees: An Introduction by Dr Nikki Gammans, Dr Richard Comont, S. C. Morgan and Gill Perkins (Bumblebee Conservation Trust, 2018) – an excellent identification guide providing all you need to know about UK bumblebees, all of which are found in Europe too.

And finally . . .

Please do support this charity helping people and bees: www.beesfordevelopment.org.

STRIPEY
CHECKLIST

These stripeys all live in my garden in South
Wales, UK. Depending on where you live, your
stripeys may be different from, though similar
to, those shown here. The point is not to know
exact scientific names, but to recognise the wide
range that you have and to ensure that they have
flowers and nesting places.

Honey bee ◯

Red-tailed bumblebee ◯

Buff-tailed bumblebee ◯

White-tailed bumblebee ◯

Garden bumblebee ◯

Tree bumblebee ◯

PLANT TREES, SOW SEEDS, SAVE THE BEES

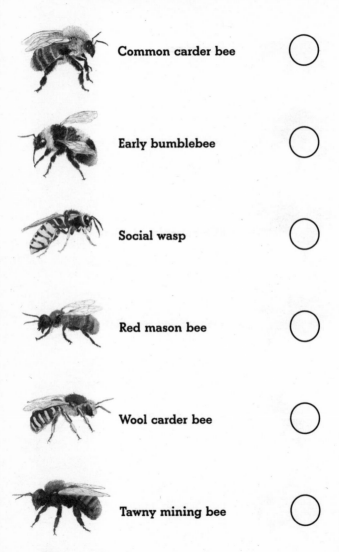

Common carder bee ◯

Early bumblebee ◯

Social wasp ◯

Red mason bee ◯

Wool carder bee ◯

Tawny mining bee ◯

Leaf-cutter bee ◯

Marmalade hoverfly ◯

Hoverfly or common drone fly ◯

Bee hoverfly ◯

Large wasp hoverfly ◯

NOTES ON STRIPEYS

PLANT TREES, SOW SEEDS, SAVE THE BEES

_____

_____

_____

_____

_____

_____

_____

_____

_____

_____

_____

_____

_____

_____

_____

_____

_____

_____

_____

_____

_____

_____

_____

_____

_____
_____
_____
_____
_____
_____
_____
_____
_____
_____
_____
_____
_____
_____
_____
_____
_____
_____
_____
_____
_____
_____
_____
_____
_____

ACKNOWLEDGEMENTS

With grateful thanks to:

My parents – Joan, who germinated my interest in plants, and Harry, who made life sweet with bees and honey;

Everyone involved with Bees for Development – from our President and Patrons, Trustees, staff, and volunteers to our beneficiaries near and far – many wonderful, kind-hearted and life-affirming people;

Kate Humble for her generous intro;

Roger Ruston and Stephanie Tyler – our local experts identifying every insect, bird and plant, and tirelessly sharing their knowledge;

All the energetic volunteers at Bee Friendly Monmouthshire;

Everyone who speaks up for wasps, cradles tired bumblebees and spends time rescuing insects from windows – armed with jam jar and postcard.